CHASING JACARANDAS

WRITTEN BY
Erin Mulay & Marybeth Ostrander

ILLUSTRATED BY **Brock Nicol**

Erin Mulay and Marybeth Ostrander
Chasing Jacarandas

TSPA The Self Publishing Agency, Inc.
Copyright © 2025 by Erin Mulay and Marybeth Ostrander

First Edition
Hardcover ISBN 979-8-9925636-0-3
Softcover ISBN 979-8-9925636-1-0

All rights reserved under International and Pan-American Copyright Conventions. Manufactured in Canada.

No part of this publication may be reproduced, stored in, or introduced into a retrieval system, transmitted in any form or by any means (electronic, mechanical, photocopying, recording, or otherwise), and/or otherwise used in any manner for purposes of training artificial intelligence technologies to generate text, including, without limitation, technologies that are capable of generating works in the same style or genre as this publication, without the prior written permission of the publisher.

This book is sold subject to the condition that it shall not, by way of trade or otherwise, be lent, resold, hired out, or otherwise circulated without the publisher's prior written consent in any form of binding, cover, or condition other than that in which it was published.

Book Design | Alicia Kowalewski
Editor | Andrea Greene
Author Portrait Photography | Jennica Maes
Publishing Management | TSPA The Self Publishing Agency, Inc.

To Audrey, Juliette, Samuel, Brielle, William, and Luca.

Thank you for discovering the Jacaranda Tree and all of its beauty. We will always cherish our Jacaranda sightseeing walks together.

The sun shines through my window.
I'm glad the rain is done.
Today is our adventure,
and walks with Mom are fun!

My cereal is crunchy.
My juice is cold and sweet.
We put hats upon our heads
and sandals on our feet.

Wind blows the door wide open,
and purple swirls around.
Our giant Jacaranda
drops petals on the ground.

Violet blooms are plenty
on the tree in our front yard.
How many spring buds are there?
To count them would be hard.

Mom says, "Let's find more trees.
We'll count them as we walk."
"Our tree is number one."
I skip on down the block.

Blossoms slick and shiny
lay on the glistening grass.
The petals make a trail
that I slide on as I pass.

A large red door leads into our village floral shop. Miss Anna's waving to me. We've come to our first stop!

"We're counting trees today, the Jacaranda kind. I love their purple flowers. How many can we find?"

The florist shows us roses.
The yellow Mom likes best.
She'll wrap them up in paper
as we go out on our quest.

Hand in hand we go,
our task to mail a letter.
"Look, Mom, four more trees!"
That's five now altogether.

The canopy above us
sways gently in the breeze.
We walk right through the tunnel
of Jacaranda trees.

A Jacaranda welcomes,
the leaves, so soft and tender.
Stop two, the postal office.
A bell rings as we enter.

Postmaster stamps our letter.
He sends it on its way.
"I've counted one more tree,"
excitedly I say!

Ahead we turn the corner.
Stop three is the pet store.
"Look, lucky number seven."
I hope we'll spot some more.

My Mom says, "Look, see that?"
A hummingbird that hovers,
busy pollinating,
sweet nectar it discovers.

We step inside the store
where birds chirp on their swings,
soft kittens snuggle tight,
and hamsters spin in rings.

"That cat would like to climb
 the Jacaranda best.
 The birds would perch on tree limbs
  and maybe make a nest."

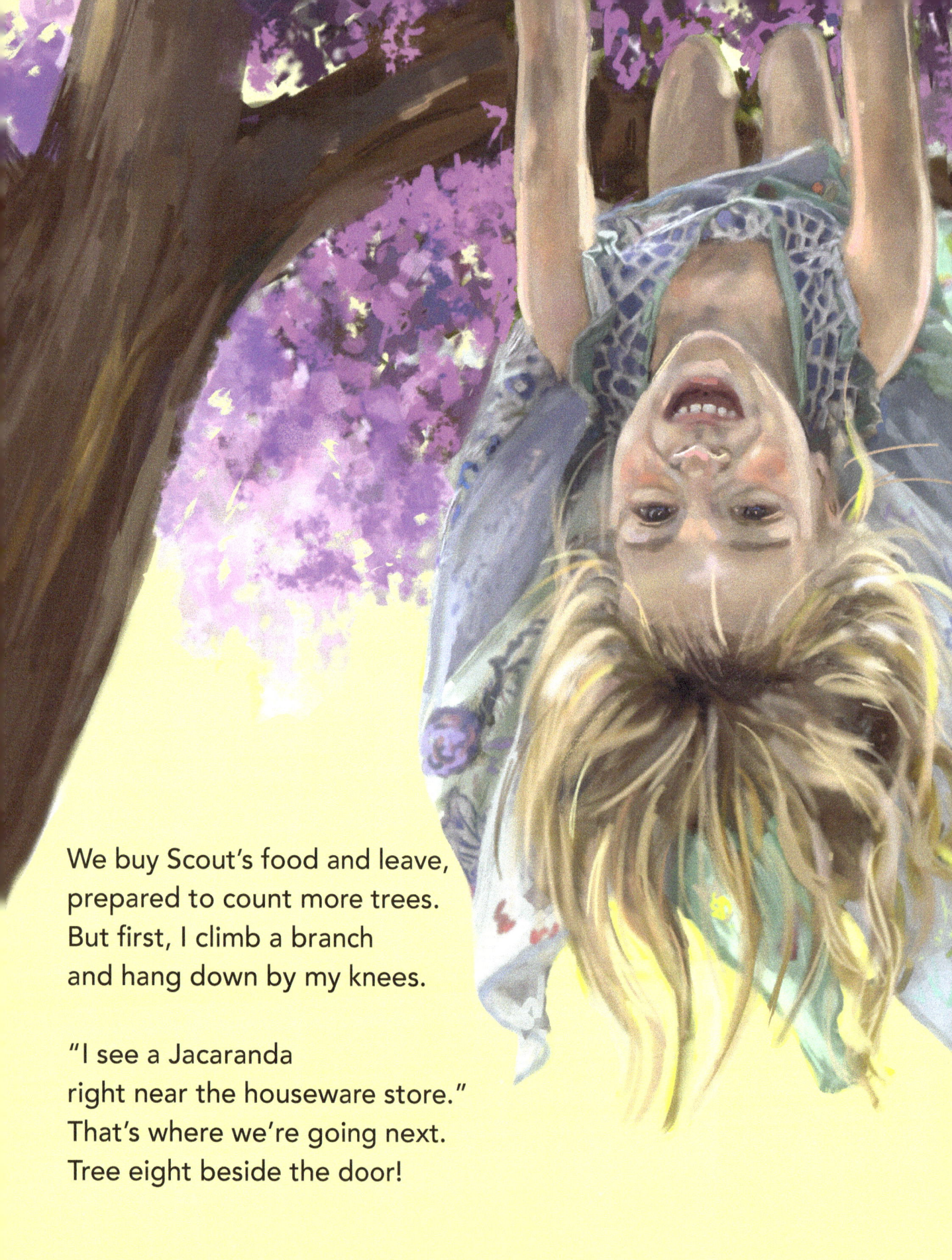

We buy Scout's food and leave,
prepared to count more trees.
But first, I climb a branch
and hang down by my knees.

"I see a Jacaranda
right near the houseware store."
That's where we're going next.
Tree eight beside the door!

We admire wooden bowls
carved from Jacaranda.
For serving food they're useful
outside on our veranda.

As we depart the store,
I touch the soft, green leaves.
The flowers feel like feathers
that brush against my sleeves.

Next door is the deli.
It's time to have our lunch.
Yummy picnic in the park,
below the trees we'll munch!

"How many trees ahead?"
"Eight more are in a group."
We lay our blanket down.
Above, the branches droop.

The breeze makes flowers dance
and petals float on air.
It's like a soft, fine rain,
gentle on my hair.

I close my eyes and listen.
Bees buzz around the park.
The shade is cool and peaceful.
"What's that?" I hear Scout's bark.

Dad's got Scout on a leash,
a pack is on his back.
He glances up and says,
"Hey, that's a fine Lilac."

"No, a Jacaranda, Dad! They bloom each spring in May. Mom and I have counted sixteen trees today!"

I lay down in the grass.
I close my eyes to rest.
I love these purple trees.
Jacarandas are the best!

GET SMART

Why is it called a Jacaranda tree?
The tree's scientific name is Jacaranda Mimosifolia or Jacaranda Acutifolia. It can also be called Black Poui or Fern Tree.

Where does it grow?
The Jacaranda tree is native to tropical and subtropical Latin America and the Caribbean. It's also found throughout Asia. It can grow in Plant Zones 9b–11. A plant zone tells gardeners what plants are most likely to grow at a certain location based on average winter temperatures. The higher the zone number, the warmer the winter temperature.

How does it grow?
In their first two years, Jacaranda trees can grow up to ten feet (3 meters) per year. After this, they grow about three feet (1 meter) per year for the next nine years. They can grow as high as 66 feet (20 meters). The flowers on the tree range from blue to indigo to violet while some trees can have white flowers. Flowers will bloom in two to three years if the tree is started from a seedling. If the tree is started from a single seed, it will take 7 to 14 years to bloom!

What is it used for?
The wood of the tree can be used for carving bowls. However, the Jacaranda's main use is ornamental. It is simply valued for its beauty.

Fun fact — A popular Christmas song in Australia called *"Christmas Where the Gum Trees Grow"* also refers to the Jacaranda tree.

Acknowledgments

We want to thank David Mulay who encouraged us to keep trying and moving forward to make our dream a reality. A big thank you to our children for igniting in our hearts a sense of wonder and joy for nature. Thank you to our editors who helped us with formatting, revision suggestions, and providing positive feedback. Thank you to Nancy Day for helping us in our early stages and struggles. Thank you to Andrea Greene for adding the final touches and to Alicia Kowalewski for building the book. Much gratitude to our illustrator Brock Nicol for bringing our characters to life on the page and showing our readers the beauty of the Jacaranda tree. Rounding off the team, thank you to the TSPA staff (Megan and Ira) for keeping the momentum going and for bringing us to the finish line.

Chasing Jacarandas is co-written by mother and daughter pharmacists who have knowledge of the medicinal and practical uses of trees and plants in our environment. As members of the Society of Children's Book Writers and Illustrators, their goal is to inspire young readers to learn more about the vegetation they encounter. This debut book is part of an ongoing series about plant life and its related careers.

Author Marybeth, a mother of two daughters and a certified teacher of young children in Montessori-based church programs, is a practicing pharmacist in her community. Her daughter Erin, a mother of six children (aged 0 to 12 years), is a member of several parenting support organizations including FIT4MOM, MOPS (Mothers of Preschoolers), Musicology, and parent groups at all school levels. She is also an active pharmacist at a compounding pharmacy, retail pharmacy, and a hospital.

www.ingramcontent.com/pod-product-compliance
Lightning Source LLC
Chambersburg PA
CBHW042357030426
42337CB00029B/5128